Monster Machines

Drilling Rigs

By Kenny Allen

Gareth Stevens
Publishing

Please visit our website, www.garethstevens.com. For a free color catalog of all our high-quality books, call toll free 1-800-542-2595 or fax 1-877-542-2596.

Library of Congress Cataloging-in-Publication Data

Allen, Kenny, 1971-
Drilling rigs / Kenny Allen.
p. cm. — (Monster machines)
Includes index.
ISBN 978-1-4339-7164-8 (pbk.)
ISBN 978-1-4339-7165-5 (6-pack)
ISBN 978-1-4339-7163-1 (library binding)
1. Oil well drilling rigs—Juvenile literature. I. Title.
TN871.2.A5357 2012
622'.3381—dc23
2011043522

First Edition

Published in 2013 by
Gareth Stevens Publishing
111 East 14th Street, Suite 349
New York, NY 10003

Designer: Daniel Hosek
Editor: Greg Roza

Photo credits: Cover, p. 1 Morkeman/Vetta/Getty Images; borders, pp. 5, 9, 13, 17, 19 Shutterstock.com; pp. 7, 21 Thinkstock; pp. 11, 15 Tyler Stableford/Stone/Getty Images.

Printed in the United States of America

CPSIA compliance information: Batch #CS12GS: For further information contact Gareth Stevens, New York, New York at 1-800-542-2595.

Contents

Boldface words appear in the glossary.

What Is a Drilling Rig?

A drilling rig is a machine that drills holes in the ground. Some are so big they need hundreds of workers. The largest drilling rigs can reach many miles underground!

The Borehole

The hole a drilling rig makes is called a borehole. Boreholes are needed to reach water to make wells. Boreholes are needed to reach oil and natural gas, too.

Get Ready to Drill!

Workers prepare the land for the drilling rig. They cut down trees. They build roads for trucks. They dig a pit where the drilling rig will be set up. They also dig a pit for waste rock.

Power Source

Drilling rigs need power to work. Some use gas **engines** that are like the motor in a car. Others run on electricity. The power **source** is set up when workers prepare the land for drilling.

The Derrick

The workers set up a derrick. This is a tall metal **frame** that holds the drill in place. It has ropes or cables that raise and lower the drill.

Drill Pipes

The derrick lowers drill pipes into the borehole. One drill pipe can be 30 feet (9.1 m) long. When one pipe is almost all the way underground, workers **connect** a new drill pipe to it. Then drilling continues.

Drill Bits

On the end of the drill pipe is a drill bit. It cuts through dirt and rock as the drill pipe goes lower. Some bits spin. Some bits have blades that cut rock. Some pound rock like a hammer.

Pump It Up!

Mud is **pumped** into the borehole through the drill pipe. The mud forces dirt and waste rock to the top of the borehole where it is removed. The dirt and rock are placed in the waste pit.

19

Monsters at Sea

An oil platform is a drilling rig at sea. Some platforms float. Others stand on the ocean floor. The tallest oil platform is 2,000 feet (610 m) tall. However, only 200 feet (61 m) of it is above water!

Oil Drilling Platforms

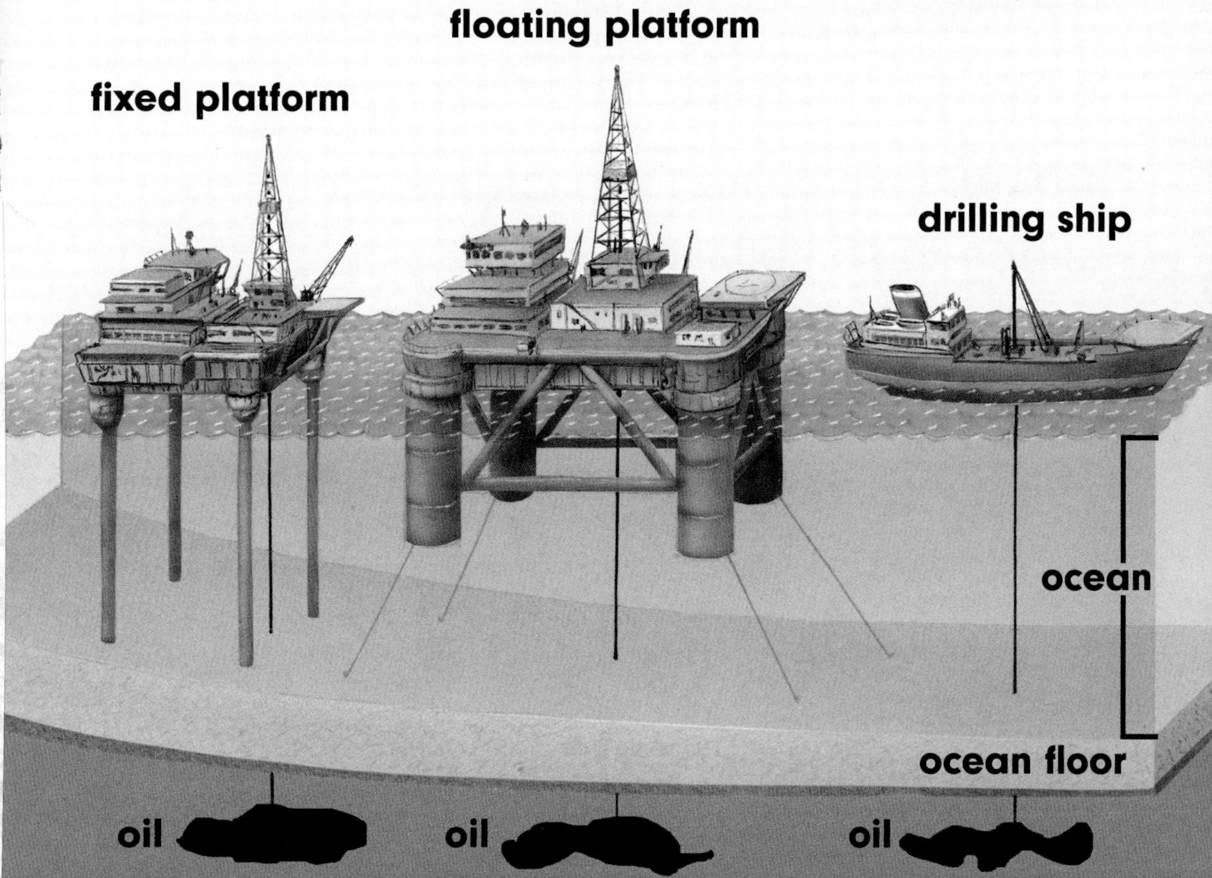

Glossary

connect: to join together

engine: a machine that makes power

frame: something that holds up or gives shape to something else

pump: to force a liquid, such as water, through a hose. Also, a machine that forces liquid through a hose.

source: the cause or starting point of something

For More Information

Books

Farndon, John. *Oil.* New York, NY: Dorling Kindersley, 2007.

Thomas, William David. *Oil Rig Worker.* New York, NY: Marshall Cavendish Benchmark, 2010.

Websites

How Oil Drilling Works
science.howstuffworks.com/environmental/energy/oil-drilling.htm
Learn more about drilling rigs and how they are used to reach oil deep underground.

Introduction to Drilling
www.pge.utexas.edu/drilling/
Take an online tour of an oil drilling rig to learn more about these monster machines.

Publisher's note to educators and parents: Our editors have carefully reviewed these websites to ensure that they are suitable for students. Many websites change frequently, however, and we cannot guarantee that a site's future contents will continue to meet our high standards of quality and educational value. Be advised that students should be closely supervised whenever they access the Internet.

Index